What Shape Is Water?

words by Nigel Croser
photographs by Martin Smith

What shape is water?

We turn on the tap, and the water comes out. The water looks like a snake!

We get a round jug.

The water runs into the jug. We look down at the water. It looks round, like the jug.

We get a square dish.

We pour the water into the dish.
Then we look down at the water.
Now it is square, like the dish.

We get a tall glass.

We pour the water into the glass. Then we look at the water. Now it is tall, like the glass.

We get a flat pan.

We pour the water into the pan. Then we look at the water. Now it is flat, like the pan.

What shape is water? Is it round or square? Is it tall or flat?

We pour out all the water, and it runs away. It does not keep a shape. It changes.

We can make water keep a shape. We pour some water in a pan with star shapes. Then we put it in the freezer.

Now the water is hard.

We have ice stars, until they melt.